D1741506

Parasitology
A Philatelic Perspective

Fred Borgsteede

Animal Sciences Group, Division of Infectious Diseases
P.O. Box 65, 8200 AB Lelystad, the Netherlands

Co-Editors

Dennis Jacobs, Emeritus Professor of Veterinary Parasitology
The Royal Veterinary College, North Mymms, Hatfield, Herts., AL9 7TA, UK

Maggie Fisher, Shernacre Enterprise, Malvern, Worcestershire, UK
Shernacre Enterprise, mfisher@globalnet.co.uk

Shernacre Enterprise
Shernacre Cottage, Lower Howsell Road
Malvern, Worcestershire, WR14 1UX, UK

First Published by Shernacre Enterprise 2007, in an edition of
1000 copies numbered 1-1000 of which this is copy:

0 1 9 3

©Shernacre Enterprise 2007

All rights reserved
No part of this publication may be reproduced, stored in retrieval system, or transmitted, in any form or
by any means, without prior permission in writing of Shernacre Enterprise.

This book is sold subject to the condition that it shall not, by way of trade or otherwise, be lent, re-sold,
hired out or otherwise circulated without the publisher's prior consent in any cover other than that in
which it is published and without a similar condition including this condition being imposed on the
subsequent purchaser.

A catalogue record for this book is available from the British Library

ISBN 0-9556488-0-9

Designed and produced by ID. Advertising, Worcester, UK.
Email: enquiries@idadvertising.co.uk

Foreword

One of the pleasurable aspects of the international conference circuit is the opportunity to wander around the historic streets of unfamiliar cities. In my case this is a somewhat aimless activity – just soaking in the ambience. Others, though, are more purposeful. I have often encountered the primary author of this book, Fred Borgsteede, in a quiet backstreet far away from the bustling activity of the shopping arcades. On occasion a loud 'Hey, Fred!' was needed to arouse him from a state of preoccupation. He was seeking out stamp dealers' shops and stalls to track down rarities missing from his philatelic collection. His schoolboy enthusiasm for exotic items from far-away places had been refined into an intellectual pursuit to record how postage stamps have been used over the years to portray parasites and parasitologists for educative, propaganda and commemorative purposes. Fred's thematic collection, if not unique, is certainly one of the finest of its kind. He is always keen to share his knowledge and enthusiasm. So when Maggie Fisher became editor of the World Association for the Advancement of Veterinary Parasitology (WAAVP) Newsletter and was seeking ideas for general interest articles, my immediate response was 'Ask Fred'. The outcome was a fascinating, informative, visually attractive and popular series. But newsletters are ephemeral, especially when distributed electronically. An opportunity to convert the series into a more tangible and durable format was identified at the 20th Conference of the WAAVP in Christchurch, New Zealand while discussing fund-raising activities for the trust fund established in memory of the late Professor Peter Nansen. The result is this delightful book. All profits from its sale will be donated to the Fund, which encourages and rewards outstanding young scientists working in the field of veterinary parasitology. I am grateful to Fred for endorsing this initiative, enhancing his original texts and adding new material, to Maggie for investing her time and considerable organisational skills, without which this project would have remained an abstract dream, and to Rita Vehbi who did much of the hard work in bringing everything together for publication. Finally, we are indebted to Merial, Novartis Animal Health Inc., Pfizer Animal Health and Central Life Sciences, whose generous sponsorship enabled us to print this book without financial risk and to maximise its fundraising potential.

Dennis Jacobs

Vice-President, WAAVP

Acknowledgements

It would have been impossible to build this collection without the help of many, many others. When so many people have helped me it is possible that I may have forgotten someone. If so, please accept my sincere apologies in the knowledge that it was not a deliberate oversight.

First of all, I would like to acknowledge my long suffering wife Wil and my children Sander, Wiedeke and Daan, because they were all 'victims' of my hobby entailing spending many hours with stamps and not with them. Secondly, Ruud Mes, who made slides of a lot of the stamps and Fred van Welie who digitalized the slides and made most of the scans. Many colleagues and friends have also helped me, particularly my fellow collector Sven Nikander and Dominique Kerboeuf, who showed me the way to Rue Drouot in Paris. The others in alphabetical order are: A. Alekseev, P. Bafort, J.-M. Doby, P. Dorchies, A. El-Harith, M. Fox, S. Frenkel, L. Gibbons, C. Gordon, J. Hansen, J. Höglund, M. Iglesias, D. Jacobs, J. Jansen, V. de Jesus Dias, V. Kharchenko, T. Kassai, D. Locke, S. Mas Coma, H. Mizgajska, R. Muller, S. Nari, C. Nielsen, D. O'Brien, D. Otranto, M. Pimentel Neto, R. Robelus, H. Schallig, J. Talmon, E. Tillman, J. Vercruysse, P. Waller and J. van Wyk.

Again my grateful thanks to those listed and to those I may have accidentally omitted to mention.

Fred Borgsteede, Lelystad, 2007.

Preface

As long as stamps have existed, people have collected them. In the 19th and the first half of the 20th century, most collectors concentrated on the stamps of their own country (and colonies if they had them). Maybe some people tried to collect the stamps of the world, but the increasing number of issued stamps and the increasing costs of such a project made this a Herculean task. In the early days the design on stamps was often an important person or a heraldic figure. Later, many other subjects were depicted and after the 2nd World War, so-called thematic philately developed. The number of themes for such specialized collections is almost infinite. Make your choice. You can collect stamps with birds, mammals, flowers, ships, Nobel Prize winners, Red Cross etc. Of course as a veterinary parasitologist, I started in 1975 to collect stamps that had a link with human or veterinary parasitology. Later on, I added stamps which depicted the scientists who had contributed to our present knowledge of parasitology. So at present my total collection stands at more than 800 stamps. How do I find these? At the start, it's a labour of love, an almost monastic task of poring through the world catalogues which list every stamp ever issued. Living on the continent of Europe, I used Michel (Germany) and Yvert et Tellier (France), because they were more accessible than Stanley Gibbons (U.K.) and Scott (USA) which certainly contain the same information. Thereafter, you have to check the philately journals which list every new issue in the world as they appear.

It is a great pleasure if you can combine your profession with your hobby, but it is an even greater pleasure if you can share this with your colleagues and all other interested persons. But it is my greatest pleasure to contribute in this way to the Peter Nansen Young Scientist Award Fund, which encourages and rewards outstanding young scientists in the field of Veterinary Parasitology.

The production of this book would not have been possible without the assistance of many others. Not only those who contributed to my collection, but particularly my Co-editors and their assistants.

Fred Borgsteede

WORLD ASSOCIATION FOR THE ADVANCEMENT OF VETERINARY PARASITOLOGY

This book was produced on behalf of the WAAVP to raise funds for the Peter Nansen Young Scientist Award Fund.

We would like to thank Merial, Novartis Animal Health Inc., Pfizer Animal Health and Central Life Sciences for their generous sponsorship without which this book would not have been produced.

Contents

CHAPTER 1

PROTOZOA: *Plasmodium*: the vector

By far the largest number of commemorative postal stamps associated with human or veterinary parasitology have malaria as their subject. The first, issued in Mexico in 1939

(1), was used in addition to the normal postal rate with the additional revenues thereby raised earmarked for the battle against malaria. The most recent was issued in France in 2004 (2), to make people aware of the continuing threat of diseases such as malaria, HIV/Aids and tuberculosis. But most stamps in this category appeared in 1962 when the World Health Organization launched a world wide campaign to eradicate malaria. No fewer than 110 countries dedicated stamps to this initiative, while others employed special slogan cancellations.

Fig. 1

Almost every aspect of the parasite itself and its vector, mosquitoes of the genus *Anopheles*, can be seen on stamps.

Fig. 2

The eggs of *A. darlingi* can be seen on a stamp from Paraguay (3). Larvae of the mosquito are found on stamps from Nigeria (4) and Tanzania (5). Adult mosquitoes are depicted on many stamps, but the most beautiful are those of the former Portuguese colonies (Angola (6), Cape Verde Islands (7), Macau (8), Mozambique (9), Portuguese Guinea (10),

Fig. 3   *Fig. 4*   *Fig. 5*

Portuguese India (11), San Tomé and Principe (12) and Timor (13). Another much chosen subject is the environment in which the mosquitoes thrive, swamp land, as seen in the stamps of Colombia (14), Cuba (15), Cyprus (16), Dubai (17), France (18), Iran (19), Lebanon (20), Monaco (21), Morocco (22) and Spain (23).

Fig. 6

Fig. 7

Fig. 8

Fig. 9

Fig. 10

Fig. 11

Fig. 12

Fig. 13

Fig. 14

Fig. 15

Fig. 16

Fig. 17

Fig. 18

Fig. 19

Fig. 20

Fig. 21

Fig. 22

Fig. 23

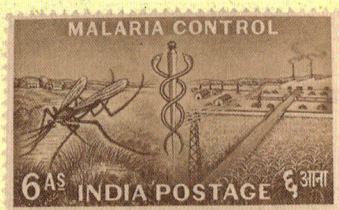

Fig. 24

Fig. 25

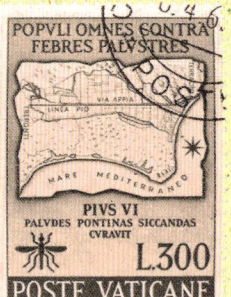

Fig. 26

Fig. 28

Fig. 27

Fig. 29

Fig. 30

Fig. 31

The relationship between swamps and the occurrence of this disease (*mal aria* = bad air) was well known long before the discovery of the life cycle of the parasite. When measures were taken to drain swamps, 'swamp fever' disappeared. The drainage of these swamp lands is depicted on stamps from India (24) and Vatican City (25, 26). Later on, when potent insecticides were developed, these were used for spraying swamps (Afghanistan (27, 28, 29), Guinea (30), Swaziland (31)). Spraying was sometimes carried out using planes (Nigeria (32), Cayman Islands (33)). Houses were also sprayed (Laos (34), San Tomé and Principe

Fig. 32

(35)), Solomon Islands (36), Turkey (37)). Many stamps simply show a spraying device but not the environment where it was used (Colombia (38), Iran (39, 40), Jamaica (41),

Fig. 33

Fig. 34

Fig. 35

Fig. 36

Fig. 37

Fig. 38

Fig. 39

Fig. 40

Fig. 41

Fig. 42

Fig. 43

Fig. 44

Fig. 45

Fig. 46

Nigeria (42), Somalia (43)), while in some cases the name of the insecticide (DDT) can clearly be seen on the spraying device (Cambodia (44), the German Democratic Republic (GDR) (45), San Tomé and Principe (46)). The stamp from South Vietnam (47) shows the

Fig. 47

Fig. 48

Fig. 49

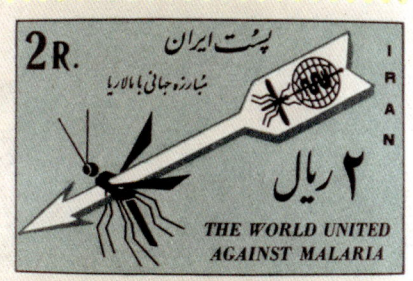

Fig. 50

Fig. 51

Fig. 52

remarkable symbolism that some countries adopted to depict the fight against mosquitoes, in this case by spraying with a hammer! Other countries have issued stamps in a similar vein. Morocco (48) and The Dominican Republic (49) portrayed a sword and Iran (50) an arrow. Tunisia heroically depicted its fight against mosquitoes by using the story of Saint George and the dragon (51) or rather more bizzarely by throttling the mosquito (52).

CHAPTER 2

PROTOZOA: *Plasmodium*: the parasite

Of course many stamps on malaria show the parasite and related matters. The life cycle of *Plasmodium* is depicted on a stamp from Kenya (53).

The diagnosis of the parasite in the blood is commonly achieved by taking a blood sample and making a blood smear, as shown on stamps from Brunei (54), Malaysia (55), San Tomé and Principe (56) and the Solomon Islands (57). Stamps from Guinea (58, 59) show investigations being performed in the laboratory. The parasite in red blood cells can be seen on stamps from Brunei (60), Cuba (61) and Poland (62, 63).

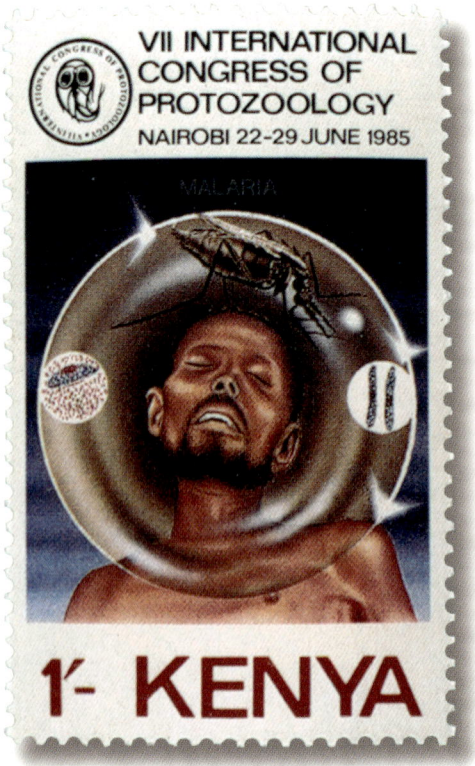

Fig. 53

Fig. 54

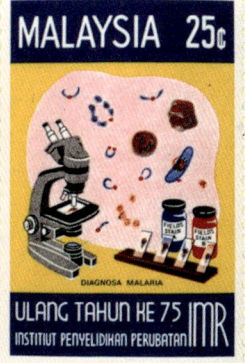

Fig. 55

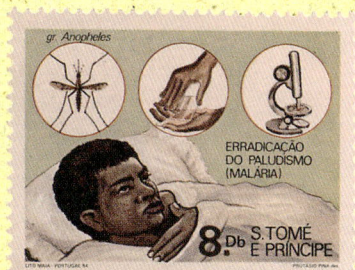

Fig. 56

Fig. 57

Fig. 58

The principle of controlling the disease by interrupting the life cycle is shown in a stamp from the Solomon Islands (64). A person suffering from malaria is depicted on stamps from the Peoples Republic of Congo (65), Kenya (53), Rwanda (66), San Tomé and Principe (56) and the Sudan (67). A simple and effective method of control is to prevent people from being bitten by mosquitoes using a mosquito net (Peoples Republic of Congo (68)).

When prevention is inadequate and people become infected, medicines have to be administered (Gabon (69), Solomon Islands (70)). A very interesting stamp with regard to anti-malarial preparations is from Nicaragua (71). It shows a bottle with the letters DARAP. The back of the stamp (visible only to the buyer and

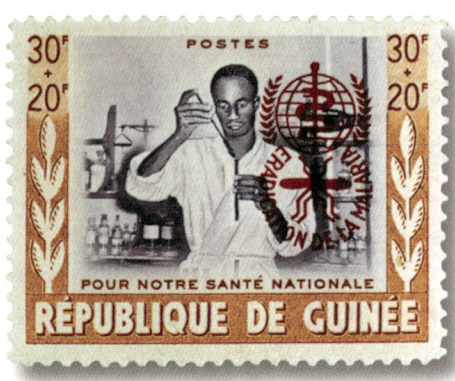

Fig. 59

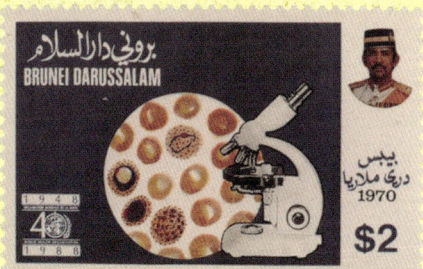

Fig. 60

Fig. 61

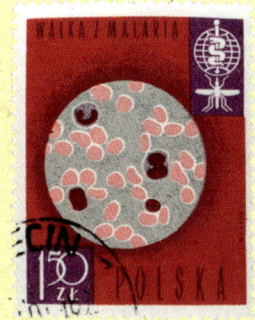

Fig. 62

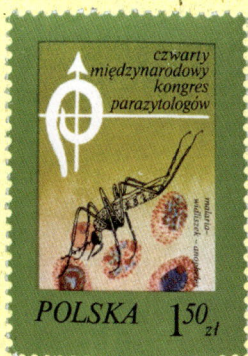

Fig. 63

stamp collector but not to the receiver of the mail) states that DARAP stands for daraprin and gives many of the facts about malaria (72).

However, the most well known medicines to treat malaria were and still are quinine and related compounds. Originally quinine came from the bark of the *Cinchona* tree (Republic of Congo (73),

Fig. 64

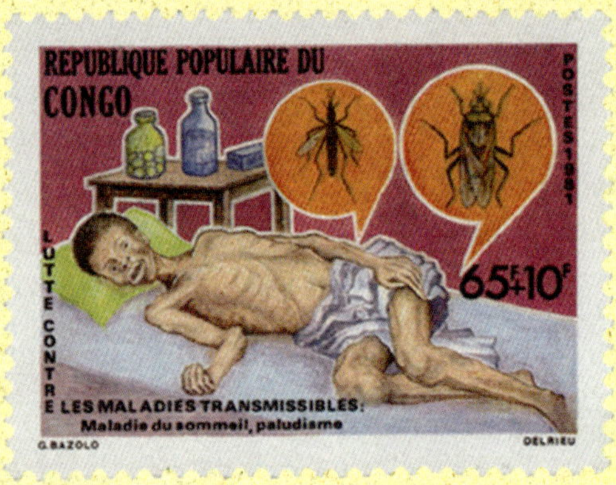

Fig. 65

Fig. 66

Fig. 67

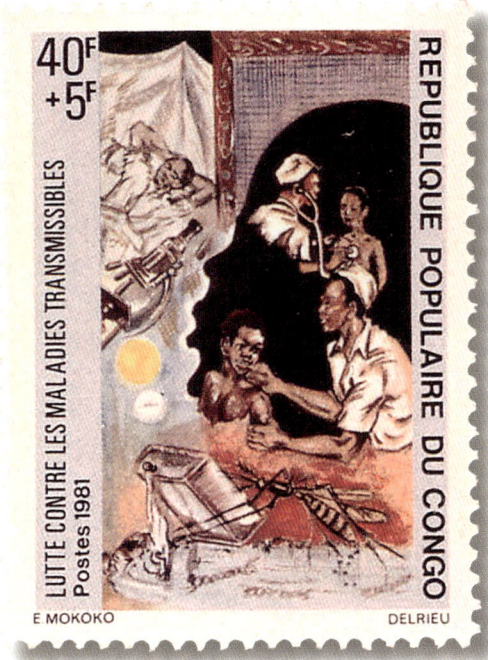

Fig. 68

Poland (74), United Nations (75), Rwanda (76, 77)). The pharmacists Pelletier and Caventou (France (78), Rwanda (79)) were the discoverers of the medicinal properties of quinine against malaria. The French stamp (78) shows the chemical structure, as does the stamp from Cuba (80).

Fig. 69

Fig. 70

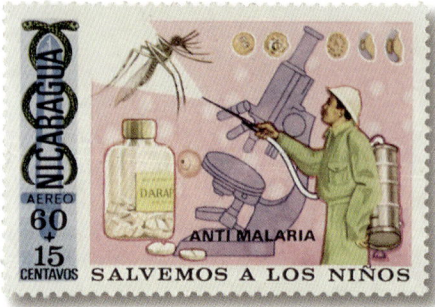
Fig. 71

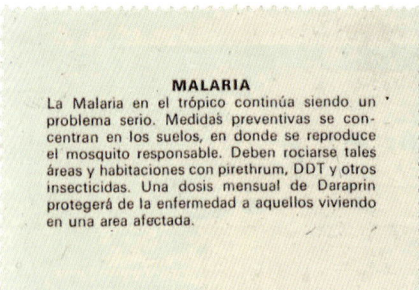
Fig. 72

Some countries were very optimistic. In 1968, Nigeria (81) issued a stamp on innoculating against malaria, while Swaziland used the word 'vaccination' as early as 1973! (82).

For the effective control of malaria, education is of the utmost importance (Laos (83)).

Indeed, the WHO campaign of 1962 made considerable progess with the number of clinical cases decreasing rapidly in the years thereafter (Israel (84)), but unfortunately the trend later reversed due to resistance - both of the vector to some commonly used insecticides and of the parasites to some of the quinine derivatives. This set-back has not been commemorated by any stamps!

Of course, many scientists have contributed to our knowledge of malaria. They are honoured on the stamps shown in Chapters 8, 9 and 10.

Fig. 73   Fig. 74

Fig. 75

Fig. 76

Fig. 77

Fig. 78

Fig. 79

Fig. 80

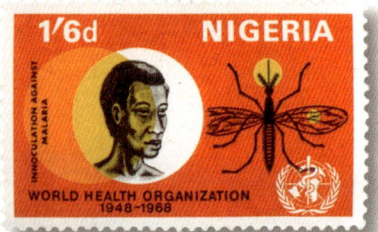

Fig. 81

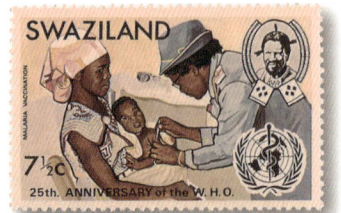

Fig. 82

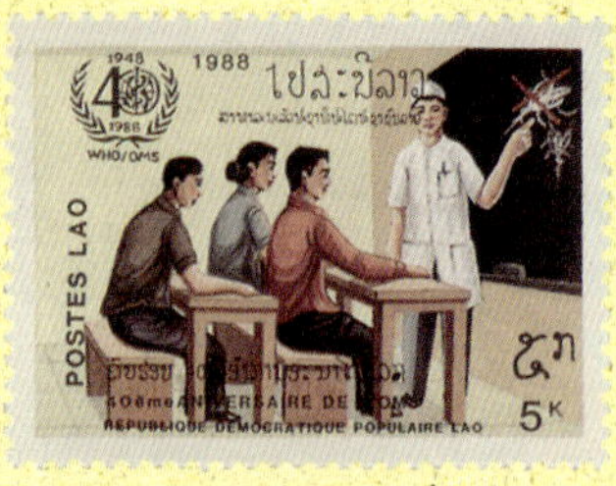

Fig. 83

Fig. 84

CHAPTER 3

PROTOZOA: *Plasmodium*: WHO campaign stamps

As noted in Chapter 1, the WHO organized an anti-malaria campaign in 1962. As far as I could discover, 110 countries participated. They have in total issued around 400 stamps. Some countries issued only one stamp, others more, but Afghanistan led with a total of 37. However, many of these differ only in value and colour and not in the depicted image.

The designs on the majority of WHO campaign stamps show one or more of the aspects of malaria that you have already seen in Chapters 1 and 2, but other stamps depict similar themes in a less striking way or just display the

Fig. 85

campaign logo. This chapter shows in alphabetical order the variety of countries that issued WHO campaign stamps in 1962 that were not displayed in the earlier chapters:

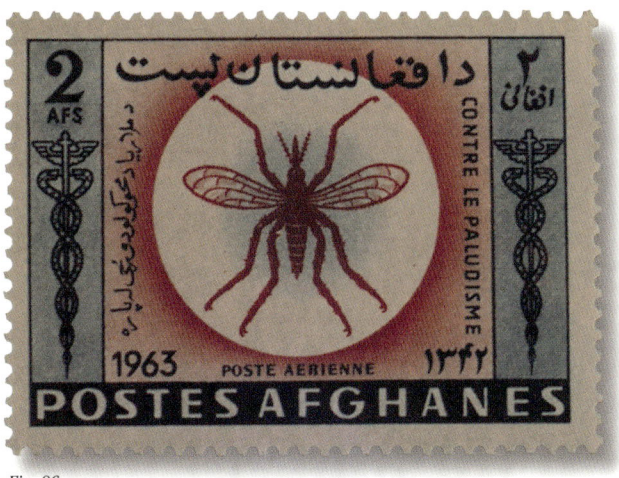

Fig. 86

Fig. 87

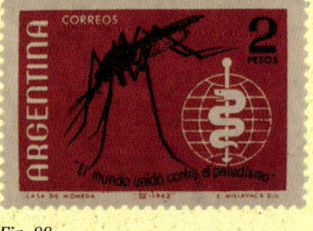

Fig. 88

Fig. 89

Fig. 90

Fig. 91

Fig. 92

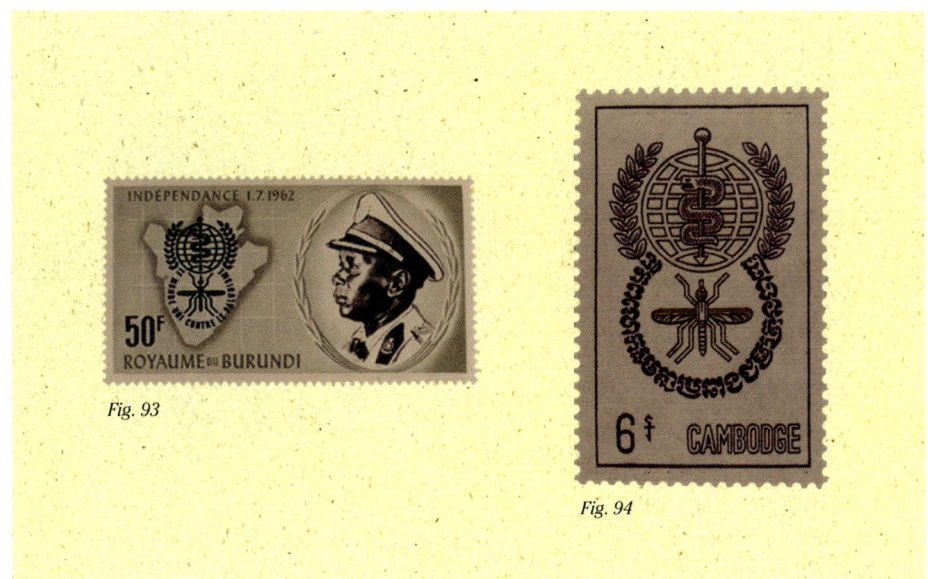

Fig. 93

Fig. 94

Afghanistan (85, 86), Albania (87), Argentina (88), Bolivia (89), Brazil (90), Brunei (91), Bulgaria (92), Burundi (93), Cambodia (94), Cameroon, representing also the other former French colonies with an identical stamp: Central African Republic, Comores, Congo, Dahomey (now Benin), Gabon, Ivory Coast, Malagasy, Mali, Mauritania, Niger, Senegal, French Somalia, Chad and Upper Volta (now Burkina Faso) (95), Ceylon (96), Republic

Fig. 95

Fig. 96

Fig. 97

Fig. 98

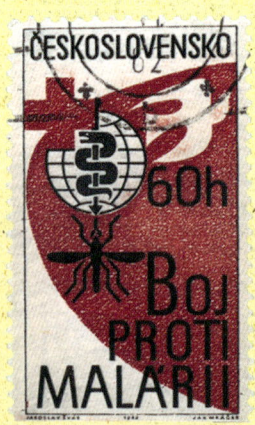

Fig. 100

Fig. 99

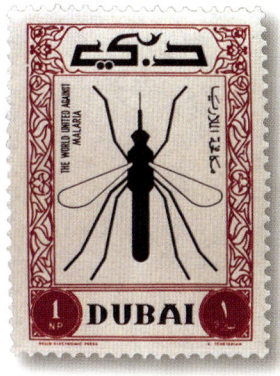

Fig. 101

Fig. 102

Congo (97), South Korea (98), Costa Rica (99), Czechoslovakia (100), Dubai (101,102), Ecuador (103), United Arab Republic (104), Ethiopia (105), Republic of China (106), German Democratic Republic (107,108), Ghana (109), Guatemala (110), Guinea (111), Haiti (112,113), Hungary (114), India (115), Indonesia (116), Iran (117, 118), Iraq (119), Italy (120), Jordan (121), Yugoslavia (122), Kuwait (123), Laos (124, 125, 126), Liberia

Fig. 103

Fig. 104

Fig. 105

Fig. 106

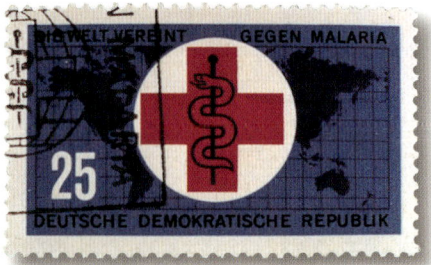

Fig. 107

Fig. 108

Fig. 109

Fig. 110

Fig. 111

(127), Libya (128), Liechtenstein (129), Malaya (130), Maldives (131), Mauritania (132), Mexico (133), Mongolia (134), Nepal (135), Nigeria (136), Pakistan (137, 138), Palestine (139), Panama (140), Panama Canal Zone (141), Papua New Guinea (142), Paraguay (143, 144, 145). Philippines (146), Poland (147), Ryu Kyu Islands (148), Saudi Arabia (149), Sjarjah (150), Sierra Leone (151), Somalia (152, 153), U.S.S.R. (154), Surinam (155), Switzerland (156), Switzerland WHO (157), Syria (158), Thailand (159, 160), Togo (161), Tunisia (162), Turkey (163), United Nations (164), United States (165), Venezuela (166), North Vietnam (167), Yemen (168,169,170, 171). Two localities whose stamps are not recognised internationally as postally valid have also issued WHO-campaign items: Herm Island (172,173,174) and Lundy (175).

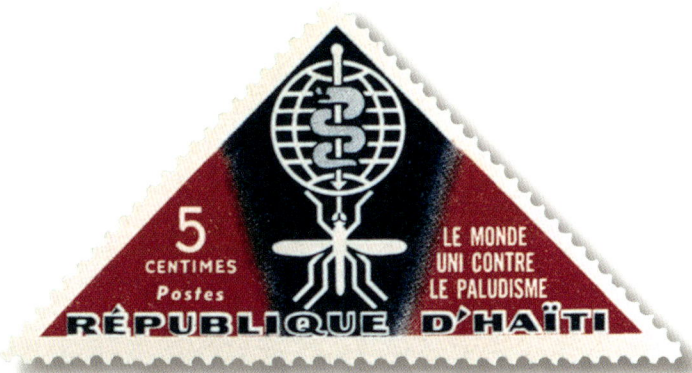

Fig. 112

Fig. 113

Fig. 114

Fig. 115

Fig. 116

Fig. 117

Fig. 118

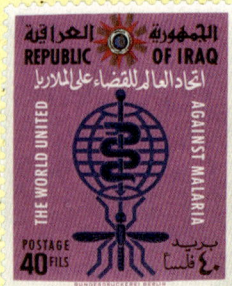

Fig. 119

Fig. 120

Fig. 121

Fig. 122

Fig. 123

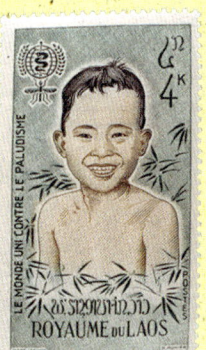

Fig. 124

Fig. 125

Fig. 126

Fig. 127

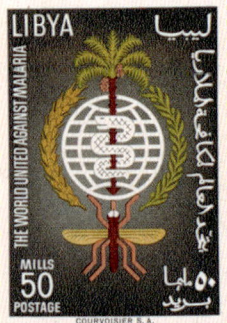

Fig. 128

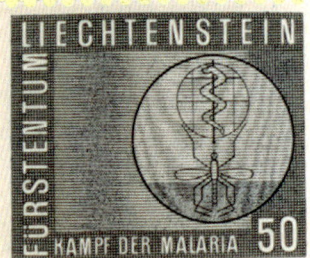

Fig. 129

Fig. 130

Fig. 131

Fig. 132

Fig. 134

Fig. 133

Fig. 135

Fig. 136

Other countries not listed so far that have issued anti-malaria stamps not in association with the WHO are: Kenya/Uganda/Tanzania (176), Indonesia (177), Republic of Maldives (178), Rwanda (179, 180) and Swaziland (181).

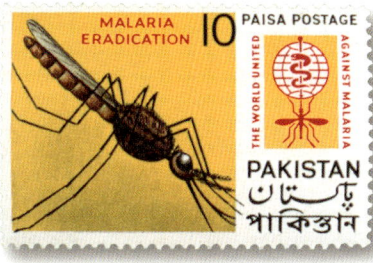

Fig. 137

Fig. 138

Fig. 139

Fig. 140

Fig. 141

Fig. 142

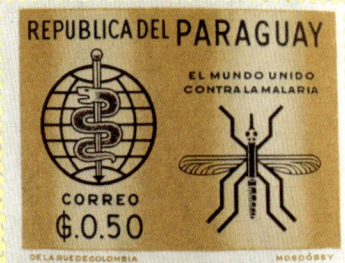

Fig. 143

Fig. 144

Fig. 145

Fig. 146

Fig. 147

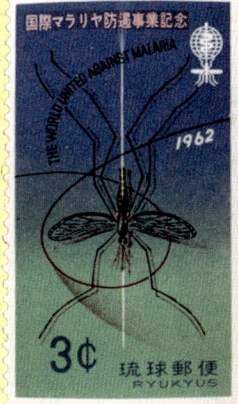

Fig. 148

Fig. 149

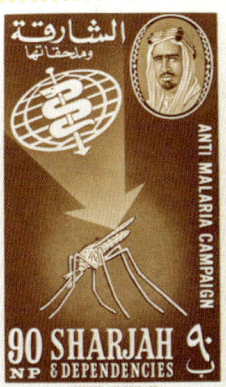

Fig. 150

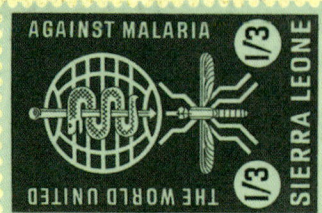

Fig. 151

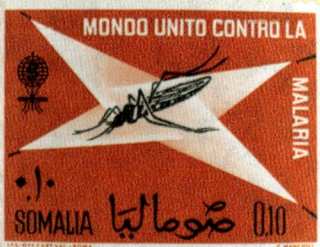

Fig. 152

Fig. 153

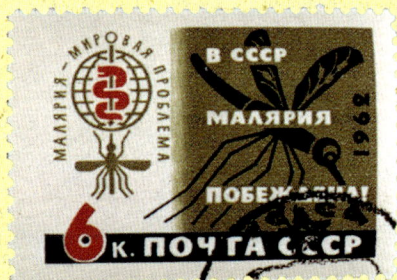

Fig. 154

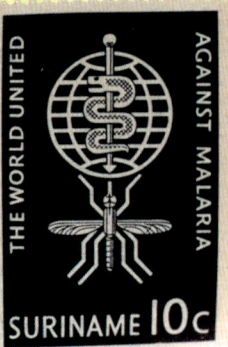

Fig. 155

Fig. 156

Fig. 157

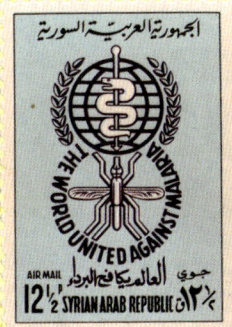

Fig. 158

Fig. 159

Fig. 160

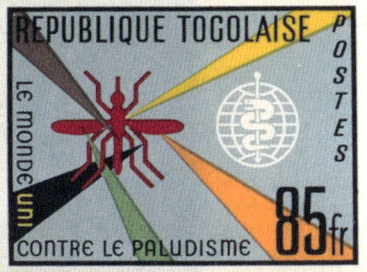

Fig. 161

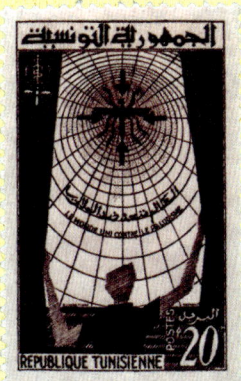

Fig. 162

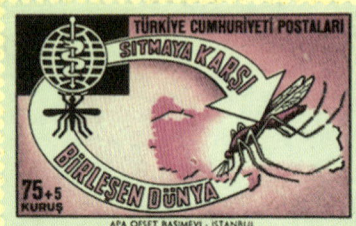

Fig. 163

Fig. 164

Fig. 165

Fig. 166

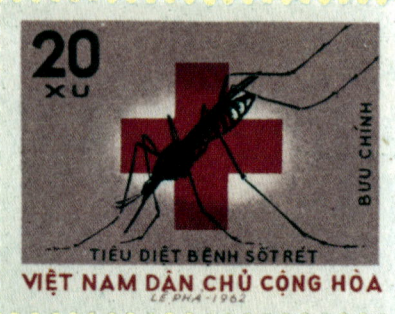

Fig. 167

Fig. 168

Fig. 169

Fig. 170

Fig. 171

Fig. 172

Fig. 173

Fig. 174

Fig. 175

Fig. 176

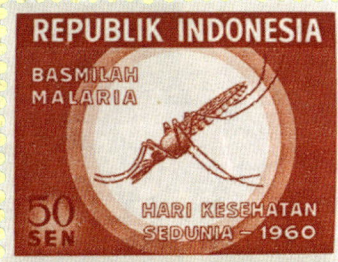

Fig. 177

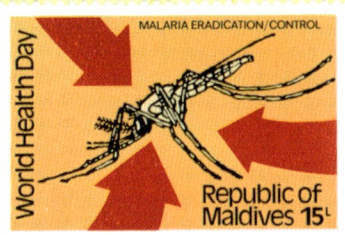

Fig. 178

Fig. 179

Fig. 180

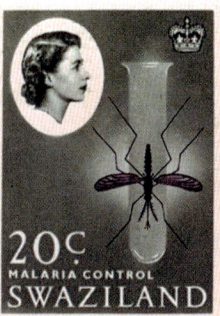

Fig. 181

CHAPTER 4

PROTOZOA: others

Besides *Plasmodium*, five other genera of protozoans can be found on stamps: *Trypanosoma, Leishmania, Babesia, Giardia* and *Theileria*.

Trypanosoma

Species of the genus *Trypanosoma* cause serious disease in man and animals in Africa and South America. The patient on the stamp from the Peoples Republic of Congo (65) is very unlucky, because it indicates that he suffers not only from malaria, but also from sleeping sickness. In certain parts of Africa, animal husbandry is only possible with trypano-tolerant cattle. The cow on the stamp from Kenya (182) with the life cycle of the parasite seems to be from a European breed. The stamp clearly shows the parasite in the blood and the tsetse fly as the vector. Vector and parasite can also be seen on stamps from the Central African Republic (183), Poland (184) and San Tomé and Principe (185). The tsetse fly is depicted on stamps of Cameroon (186, 187), Gabon (188), Kenya (189), Tanzania (190) and Upper Volta (191). A stamp from Brazil is dedicated to the South American form of sleeping sickness, 'Chagas' disease (192). *Trypanosoma cruzi*, the cause of this disease, is transmitted by bugs. Maybe it is the 'kissing bug', *Triatoma infestans*, in the lower right corner?

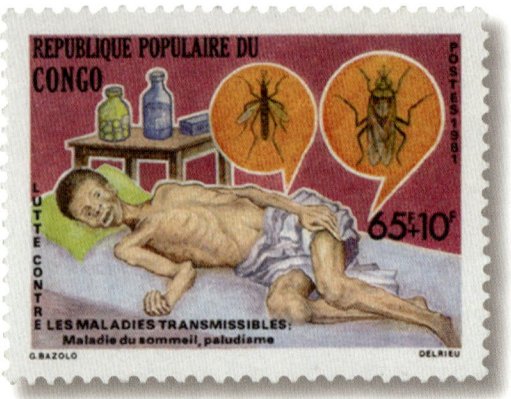

Fig. 65

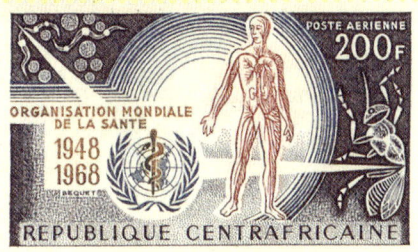

Fig. 183

5/- KENYA

Fig. 182

Fig. 184

Fig. 185

Fig. 186

Fig. 187

Fig. 188

Fig. 189

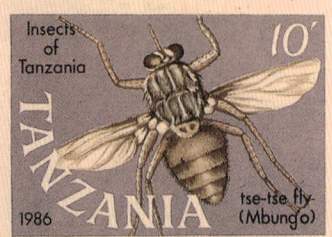

Fig. 190

Fig. 191

Leishmania

The life cycle in man is shown on a stamp from Kenya (193), while the life cycle in the Syrian hamster can be found on a stamp from Israel (194). A blood smear may be one way to detect the parasite (Brazil (195)). Money for research into a vaccine against *Leishmania* is the subject of another stamp from Brazil (196).

Fig. 192

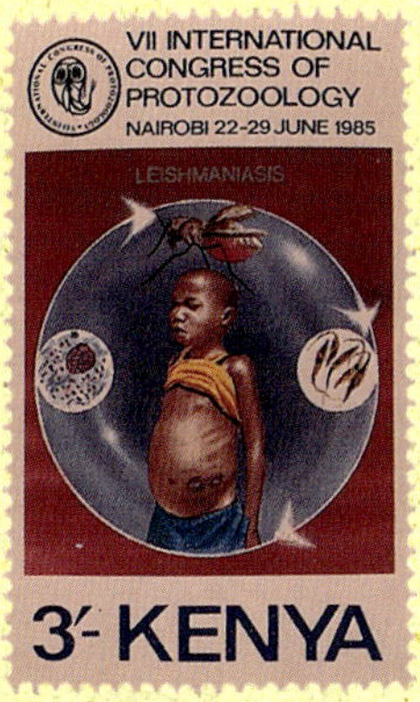

Fig. 193

Fig. 194

Fig. 195

Fig. 196

49

Fig. 197

Fig. 198

Fig. 200

Fig. 199

Fig. 201

Fig. 202

Babesia

Again, a stamp from Kenya shows the life cycle of *Babesia*, in this case in dogs (197). Now, we have seen all four stamps of this series (53, 182, 193,197). There is something worth noting about the wording on these stamps as all the parasite genera involved end with the letter 'a', yet the names of two of the associated diseases end with '-iasis' and one with '-osis'. According to the internationally standardised convention for the nomenclature of parasitic diseases, SNOPAD, their correct names should be 'leishmaniosis' and 'trypanosomosis', while babesiosis is correct. The postal authority must have been advised by a parasitologist of the old anarchic tradition! The disease babesiosis is named after the famous Romanian scientist Babes. His homeland honoured him with many stamps (198, 199, 200, 201).

Giardia

Giardia can be seen in the upper left hand corner of the series from Kenya now well known to the reader (53, 182, 193, 197). Although this book is on stamps and not on cancellations, I would like to make one exception for a special cancellation that was used during a protozoology congress in Leningrad, USSR, the present St. Petersburg in Russia (202).

Theileria

There is only one item to be found under this heading, a stamp dedicated to Sir Arnold Theiler, a famous parasitologist and founder of the Onderstepoort Veterinary Institute in Pretoria, South Africa. The stamp shows his portrait and the parasite in the blood (203).

Fig. 203

CHAPTER 5

Cestodes and trematodes

Fig. 204

Would you be surprised by the fact that the Platyhelminthes are not well represented on stamps? Perhaps the diseases they cause do not claim as many victims as diseases caused by protozoans, but they are there!

Cestodes

Echinococcosis is the subject of two stamps. The first one is from Algeria (204). It shows the home of the adult tapeworm (intestine of the dog) and the places in humans where cysts can occur (liver, lungs and brain).

The second stamp is from Uruguay (205). This is dedicated to a national campaign to control echinococcosis by preventing dogs from feeding on offal harbouring the larval cestodes (hydatid cysts).

Fig. 205

Fig. 206

Fig. 207

Fig. 208

Trematodes

Of course, in man the most important trematode genus causing disease is *Schistosoma*. The oldest stamp depicting a schistosome pair and egg, in this case *Schistosoma mansoni*, is from Brazil (206). The United Arab Republic honours Theodor Bilharz who, in 1852, discovered schistosomes in the veins of an Egyptian patient (207). Another schistosome pair is depicted on a stamp from Egypt (208).

In 2000, Cuba issued a stamp on which *Fasciola hepatica*, the liverfluke, can be seen (209). This is a parasite that causes huge economic losses in cattle and sheep. However, the parasite also infects humans, particularly in some countries of South America, such as Bolivia and Peru. High in the Andes, up to 80% of the inhabitants of villages can be infected. The stamp is dedicated to a scientist, Dr. Pedro Kourí Esmeja, who has studied human aspects of this disease.

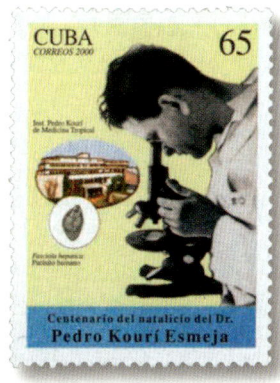

Fig. 209

53

CHAPTER 6

Nematodes

Fig. 210

Nematodes are much better represented on stamps than trematodes or cestodes. They are the cause of important tropical diseases such as river blindness and dracunculosis, 'guinea worm disease'. But nematodes are not only harmful in humans, but also in livestock where they cause huge economic losses.

Fig. 211

Onchocercosis

Onchocercosis, river blindness, is common in West African countries, such as Niger, Mali, Ghana, Togo and Burkina Faso (former Upper Volta). Several aspects of the disease can be seen on stamps. A blind person is depicted on stamps from Togo (210). A stamp from Upper Volta shows a young boy guiding

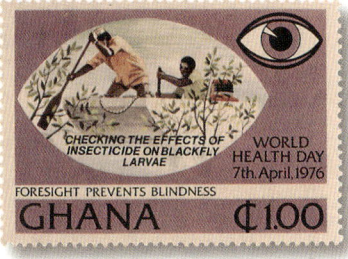

Fig. 212

Fig. 213

Fig. 214

three blind women (211). Investigation of the eye is found on stamps from Ghana (212, 213) and Togo (214). Control of the simuliid blackflies, vectors of the worm, is depicted on stamps from Ghana (215) and Upper Volta (211). Other stamps that refer to the battle against onchocercosis are from Ghana (216), Niger (217) and Upper Volta (218).

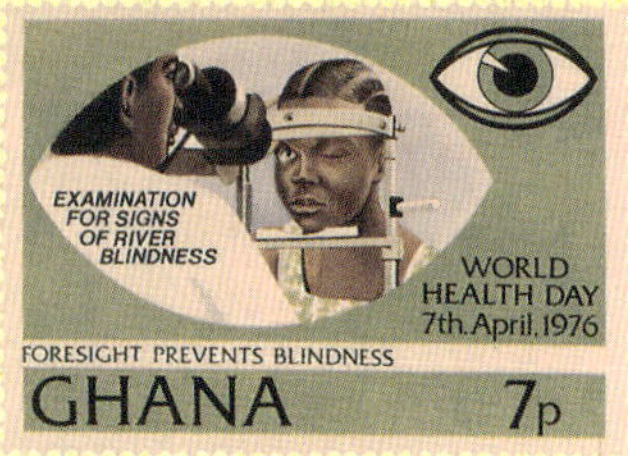

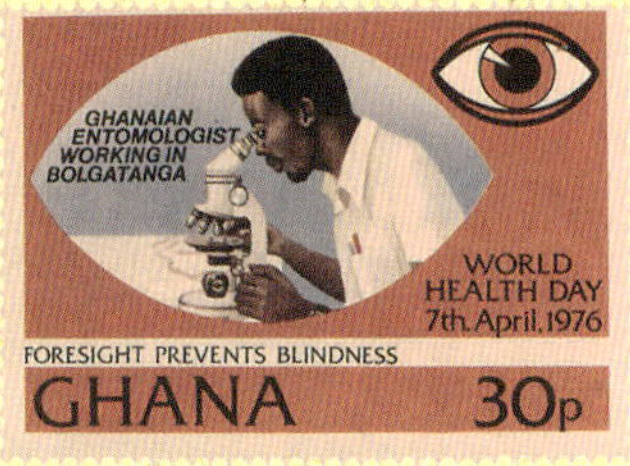

Fig. 215

Dracunculosis

Dracunculus medinensis, the guinea worm, is now eliminated from most tropical countries. Eradication campaigns in Nigeria and Sudan motivated the issue of a series of stamps. The Nigerian stamps show the swollen leg often associated with this disease (219), the places where people can pick up the infection from drinking water containing the copepod intermediate host (220) and the way to prevent infection by boiling drinking water (221). Sudanese stamps show a foot where the worm has penetrated the skin to eject larvae from the body (222) and

Fig. 216

Fig. 217

Fig. 218

Fig. 219

Fig. 220

Fig. 221

an (infected?) boy (223). The third stamp is dedicated to a meeting on the eradication of the guinea worm held in Khartoum in 2002 (224).

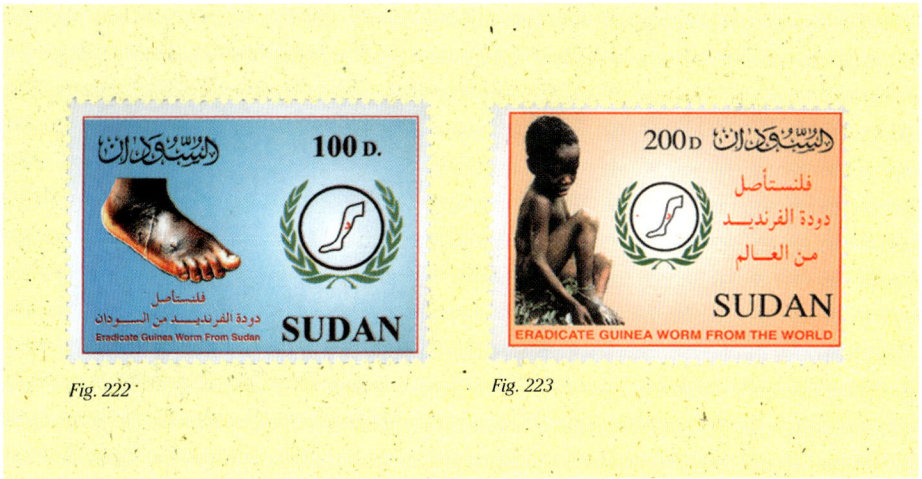

Fig. 222

Fig. 223

Nematodes of veterinary importance

Examples are displayed in an interesting series from Cuba issued in 1975. The stamps show the host and the life cycle of the parasite. Thus, we can find *Haemonchus* spp. in sheep (225), *Ancylostoma caninum* in the dog (226), *Dispharynx nasuta* in poultry (227) and '*Ascaris lumbricoides*' in pigs (228), although the latter is more likely to be *Ascaris suum*!

Fig. 224

Fig. 225

Fig. 226

Fig. 227

Fig. 228

Control of worms

The Caribbean island of St. Vincent has issued a block of stamps showing medicinal plants. Among these, there are two with a proven effect against 'worms': pussley (*Portulaca oleracea*) (229) and guava (*Psidium guajava*) (230).

Fig. 229

Fig. 230

CHAPTER 7.

Arthropods

We saw in Chapters 1, 3, 4 and 6, that there are many stamps showing the insect vectors of parasites: species of the genera *Anopheles, Glossina, Phlebotomum, Simulium* and *Triatoma*. But, also 'real' arthropod parasites can be found on stamps. The majority are ticks, but there are also some insects.....

Ticks

By far the most impressive series on ticks has been that issued by Mozambique in 1980. The six stamps show respectively *Dermacentor circumguttatus cunhasilvai* from the elephant (231), *Amblyomma hebraeum* from the giraffe (232), *Dermacentor rhinocerinus*

Fig. 231

Fig. 232

from the rhino (233), *Amblyomma pomposum* from a gazelle (234), *Amblyomma theilerae* from a cow (235) and *Amblyomma eburnum* from a buffalo (236). A stamp from Cuba illustrates the three-host life cycle of the cattle tick *Boophilus microplus* (237). The tick as vector of disease, in this case scrub typhus, is depicted on a stamp from Malaysia (238). The stamp also gives the drug used to cure this disease: chloramphenicol. Besides these, ticks can also be seen on stamps from Kenya (197, 239). The latter (239) is one of a series issued to celebrate the 25th anniversary of the ICIPE (International Centre of Insect Physiology and Ecology) which employs a tick as its symbol - quite remarkable for an organisation with 'insect' in its name. Maybe a better acronym would be ICAPE, where the 'I' of 'insects' has been replaced by the 'A' of 'arthropod'.

Fig. 233

Fig. 234

Fig. 235

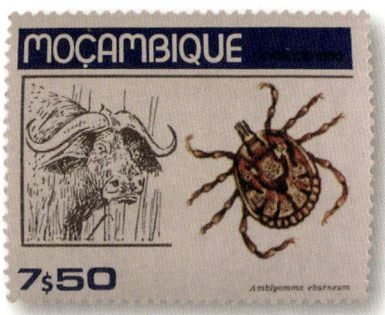

Fig. 236

Mites

Mites on stamps are rare. There is a stamp from Chad showing *Dinothrobium tinctorium* (240) which belongs to the so-called velvet mites. In the past, these were the source of a bright red dye. In their larval stage, these mites are parasitic on (i.e. ectoparasites of) insects and other invertebrates.

Another stamp that can be associated with mites comes from Italy (241). It is dedicated to the historical Latin writer Varro (116 – 27 B.C.) who described the way the Romans kept bees for honey production. He was given the dubious 'honour' of having the genus *Varroa* named after him. *Varroa jacobsonii* is the notorious and economically damaging honeybee mite that can decimate bee-hives.

Fleas

Believe it or not, there is a stamp showing a flea (242). This is on a drawing by a German child issued in the series for youth in 1971. And although it is difficult to identify the species, it is highly probable that the flea represents *Ctenocephalides felis*, by far the commonest flea on dogs and cats in northwestern Europe.

Fig. 237

Fig. 238

Bots

The horse bot, *Gasterophilus intestinalis*, and its life cycle can be found on one of the stamps of the famous veterinary parasitology series from Cuba (243).

Fig. 239

Fig. 240

Fig. 241

Fig. 242

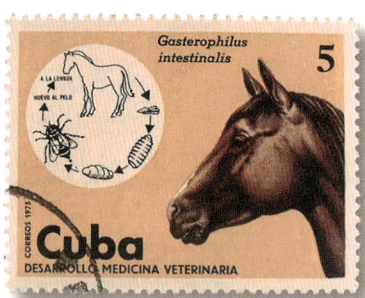

Fig. 243

Fig. 244

Myiasis

Lucilia sericata is the most important myiasis-causing fly in the temperate climatic zone of Europe. The fly is depicted on a stamp from Ascension Island, issued in 1989 (244).

Fig. 245   Fig. 246

Control of ectoparasitic infections

Insecticides are commonly used to control ectoparasitic infections on animals and man. One group of insecticides is based on pyrethrum, originally extracted from plants of the genus *Chrysanthemum*, as shown on a stamp from Rwanda (245). Sheep can be treated by dipping, as can be seen on a stamp from the Falkland Islands (246). Although dipping is an effective way of treatment, in many countries it is no longer allowed due to regulations aimed at preventing pollution to the environment.

CHAPTER 8

Scientists who made major contributions to parasitology (from B.C. – 1800 A.D.)

Fig. 247

Our present state of knowledge would not have been achieved but for the curiousity of people in the past who were fascinated by the creatures that live within and on the surface of animals and man.

In this chapter, we start with three famous historical generalists. They were interested in many aspects of nature, the human body and matters related to the body. They certainly deserve to be honoured with a stamp.

The founders of human medicine (from B.C. – 1100 A.D)

Hippocrates (460 - 370 B.C.) on stamps from Greece (247), Hungary (248), San Marino

64

Fig. 248

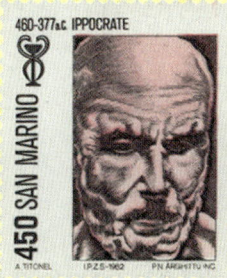

Fig. 249

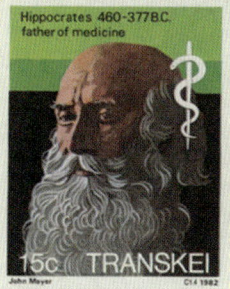

Fig. 250

Fig. 251

Fig. 252

Fig. 253

(249) and the Traanskei (250) already knew of the roundworm, tapeworm and pin-worm of man. **Aristotle** (384-322 B.C.) mentioned animal parasites, such as roundworms in horses and cysticerci in swine, of course without knowing anything about their life cycles. We can see him on stamps from Greece (251, 252) and Cyprus (253). Besides Hippocrates, **Avicenna (Ibn Sina)** (980 -1037 A.D.) can be regarded as a great philosopher and the founder of human medicine. He was familiar with a lot of parasitic diseases in humans and many Arabic countries have devoted stamps to him. Because there are so many, we have chosen those from Algeria (254), Dubai (255), Kuwait (256) and Syria (257). Iran has issued a stamp with both Avicenna and Hippocrates (258).

From 1100 – 1800 A.D.

In these centuries, science was still more or less 'general'. **Hildegard von Bingen** (1098 – 1179) is perhaps better known for her religious works and music, but she also wrote in 'Physica' a treatise on intestinal worms and she gave recipes for the treatment of worms in dogs. She can be seen on a stamp from Germany (259). Contributions to parasitology also came from **Albertus Magnus** (1193 - 1280), who described intestinal worms of horses, dogs, falcons and fish. Only his home country Germany has issued stamps dedicated to him (260, 261, 262).

Fig. 254

Fig. 255

Girolamo Savonarola (1452 -1498), more famous as being the instigator of the 'bonfire of the vanities', wrote in his 'Tractatus de vermibus' about intestinal worms in humans (263).

Theophrastus Bombastus von Hohenheim, better known as **Paracelsus** (1493 -1541) introduced new theories

66

about helminth life cycles and gave instructions for treatment. He is depicted on stamps from Germany (264), Hungary (265) and Switzerland (266).

Andreas Vesalius (1514 -1564) is well known for his anatomical studies. But during his dissections, he also found helminths and wrote about them. Stamps are from his home country Belgium (267, 268), as well as from Hungary (269) and Traanskei (270). Together with the Arabic scientist Ibn Khaldoun, he is on a stamp from Tunisia (271).

Fig. 256

Fig. 257

Fig. 258

Fig. 259

The 'father' of microscopy is **Antonie van Leeuwenhoek** (1632 – 1723). Under his microscope he observed protozoans, nematodes and acanthocephalans. He is depicted on stamps from the Netherlands (272) and Traanskei (273).

Marcello Malpighi (1628 -1694) discovered '*Cysticercus cellulosae*' in many mammals and birds. He can be found on stamps from Italy (274) and Traanskei (275).

Fig. 260

Fig. 261

Fig. 262   *Fig. 263*   *Fig. 264*

Hermanus Boerhaave (1668 -1738) was a leading medical doctor in his time. He was the first who suggested a relationship between eating fish and tapeworms. The Netherlands dedicated two stamps to him (276, 277).

Fig. 265   *Fig. 266*

One cannot forget **Carl von Linné** (Linnaeus) (1707-1778). He wrote his famous 'Systema naturae' in Leiden, the Netherlands, where he studied under Boerhaave. In Part 6 he mentioned 'Vermes', but, in fact, it had not much to do with parasitology. In his 12th edition of the 'Systema naturae', *Ascaris* and *Fasciola* were the only real parasites among

Fig. 267

Fig. 268

Fig. 269

Fig. 270

Fig. 271

many other invertebrates. He appears on several stamps. Here are examples from the German Democratic Republic (278), San Marino (279), Soviet Union (280) and Sweden (281, 282). In 2007, Sweden issued a stamp on the 300th anniversary of his birth (283).

Fig. 272   *Fig. 273*

Fig. 274

Fig. 275

Fig. 276

Fig. 277

Fig. 278

Fig. 279

Fig. 280

Fig. 281

Fig. 282

Finally, **Peter Christian Abildgaard** (1740 -1801), the founder of the Royal Veterinary and Agricultural University in Copenhagen, was the first 'real' parasitologist. He not only worked on helminth systematics, but he was also the first to carry out experimental infection studies with tapeworms. It is of course Denmark that honours him with a stamp (284).

Fig. 283

Fig. 284

CHAPTER 9

Scientists who made major contributions to parasitology (after 1800 A.D.)

Here we mention in chronological order scientists of the past two centuries who made major contributions to the advancement of parasitology. I realise that this is a subjective choice and if your favourite is not here, you may perhaps find the missing name in the next chapter or maybe no stamp has been issued as yet.

Fig. 285

Fig. 286

Fig. 287

Fig. 288

Pierre-Joseph Pelletier (1788 – 1842) and **Joseph-Bienamé Caventou** (1795 -1842) isolated in 1820 an alkaloid from the bark of the quina tree from Peru which they called quinine. After medical experiments by others, it was established that this compound was indeed active against malaria. They set up a factory in Paris for extraction of quinine and since then the purified compound has been used instead of the powdered bark to treat malaria. Their stamps were mentioned earlier (78, 79).

Rudolph Ludwig Carl Virchow (1821 – 1902) was a famous German pathologist and later politician. He wrote about the pathology of helminth infections in man, particularly *Trichinella, Ascaris, Enterobius, Echinococcus granulosus* and *E. multilocularis*. He can be seen on stamps from Germany (285, 286, 287) and Hungary (288).

Theodor Bilharz (1825 – 1862) was the person who discovered in 1852 the causative parasite of a common disease among people in Egypt. The disease was called after him: bilharziosis, now mostly named schistosomosis. Stamp 207 is dedicated to him.

Fig. 289

Fig. 290

Robert Koch (1843 – 1910) was mainly a bacteriologist. However, he also spent some time in Italy and Africa, where he worked on malaria, *Trypanosoma*, *Babesia* and other tick-transmitted diseases. Many stamps depicting Koch have been issued. Here we show just one, from Germany (289). He received the Nobel Prize in Physiology or Medicine in 1905 (290).

Camillo Golgi (1843 – 1926) is well known for the methods he developed to stain nerves and cell structures (291). Working in Pavia, Italy, he became interested in malaria. In 1890, he found a way of photographing the characteristic phases of the parasite. He was awarded the Nobel Prize in Physiology or Medicine in 1906 (292).

Fig. 292

Fig. 291

Fig. 293

Fig. 294

Charles Louis Alphonse Laveran (1845 – 1922) was the first to discover the blood stadia of *Plasmodium* in a patient from Algeria. His first papers were received with a good deal of scepticism, but gradually his colleagues accepted his findings. Algeria honoured him with a stamp (293). He was awarded the Nobel Prize in Physiology or Medicine in 1907 (294).

Elie Metchnikoff (1845 – 1916) was famous for his work on phagocytosis and received the Nobel Prize in Physiology or Medicine in 1908 (295). He was born in Russia, but worked mainly in Germany, Russia, Italy, Austria and France. As a student, he collaborated with Karl Leuckart in Giessen, Germany, on the life cycle of *Rhabdias bufonis*, the lung nematode of toads and frogs. He features on stamps from France and several from the USSR (296, 297, 298, 299).

Fig. 295

Fig. 296

Fig. 297

Fig. 298

Fig. 299

Giovanni Battista Grassi (1825 – 1925) was convinced that mosquitoes transferred malaria and he recognised the *Anopheles*-type as the true carrier, unaware of the work of Ronald Ross. Italy dedicated a stamp to him (300).

Victor Babes (1854 – 1926) gave his name to the genus Babesia. In fact, he was a microbiologist working in Bucharest, Romania. He discovered more than 50 new bacteria and initiated serotherapy. Romania honoured its famous son with many stamps (198, 199, 200, 201).

Fig. 300

Fig. 301

Adolfo Lutz (1855 – 1940) was a Brazilian physician who is regarded as the father of tropical medicine and medical zoology in Brazil. His work on yellow fever and transmission by *Aedes aegyptii* found international recognition. His name lives on in *Anopheles lutzii* and the *Dicrocoelium*-like trematode genus *Lutztrema*. He can be seen on a stamp from Brazil (301).

David Bruce (1855 – 1931) discovered the agent causing 'nagana', sleeping sickness. Later, the agent was named after him, *Trypanosoma brucei*, although the bacterial genus *Brucella* is probably more well known. Malta has a stamp with Bruce on it (302).

Julius Wagner-Jauregg (1857 – 1940) was an Austrian physician who received the Nobel Prize in Physiology or Medicine in1927 for his work on the cure of mental diseases

Fig. 302

Fig. 303

Fig. 304 · Fig. 305

by means of inducing fever. He did studies with tuberculin, but without much result. So he turned to malaria inoculation in patients with 'dementia paralytica'. They proved to be very successful. Would this type of work meet the ethical standards of today? He is on a stamp from Austria (303).

Ronald Ross (1857 – 1932) was able to demonstrate that the theories of Laveran and Manson concerning the transmission of malaria by mosquitoes were correct. After this, he devoted almost all of his life to the study of malaria and the means of controlling the disease. Is it no wonder that he was awarded the Nobel Prize in Physiology or Medicine in 1902 (304). Because his experimental investigations testing Laveran's theory were done in India, this country dedicated a stamp to him (305).

Charles Jules Henry Nicolle (1866 – 1936) studied medicine in France and worked on several subjects such as cancer, transmission of typhus by the body louse, tick fever and the cultivation of *Leishmania* in artificial media. For his work, he received the Nobel Prize in Physiology or Medicine (306).

Fig. 306

Arnold Theiler (1867 – 1936), born in Switzerland, and moved after his studies in Bern and Zürich in 1891 to South Africa. He published many papers on all kinds of tropical diseases, including those concerned with parasites. He was the first to discover that East Coast Fever was caused by a protozoan in blood that was transmitted by ticks. Therefore the genus *Theileria* was named after him (203).

Fig. 308

Fig. 307

Jules Bordet (1870 – 1961) gave his name to the causative agent of whooping cough: *Bordetella*. Coming from Belgium, he worked at the famous Pasteur Institute in Paris. Later, he returned to Brussels and founded the Pasteur Institute in that city. His immunological work included studies on immunity against malaria. He was rewarded with the Nobel Prize in Physiology or Medicine in 1919 (307, 308).

Oswaldo Goncalvez Cruz (1872 – 1917), born in Brazil, studied medicine and worked on yellow fever, malaria and Chagas' disease. The trypanosome causing this disease was named after him. But his name cannot only be found as a species name, a whole genus is named after him too, the nematode genus *Oswaldocruzia*. Oswaldo Cruz can be found on two stamps from Brazil (309, 310).

Fig. 310

Fig. 309

81

Fig. 311

Manuel Augusto Piraja da Silva (1873 – 1961) was a pioneer in the field of microscopical identification of intestinal schistosomosis, amoebic dysentery, cutaneous leishmaniosis, Chagas' disease and many others. In 1908, he gave a complete description of *Schistosoma mansoni* (206).

Konstantin Ivanovitch Skrjabin (1878 – 1972) must be a familiar name for every parasitologist that does biosystematic work on helminths. He is the first author of that unique series on the systematics of nematodes and trematodes, originally only in Russian, but also translated into English. Although there is no stamp with Skrjabin on it, there is an item of postal stationery carrying his portrait (311)!

George Gigliolo (1879 -1975) is perhaps lesser known internationally, but he played a very important role locally in the fight against malaria in Guyana. In 1946 he started large scale control measures against mosquitoes by means of DDT spraying. In 1951, the coastal areas were almost free of *Anopheles darlingi*. Later, in the mid-1960s, by distributing anti-malarials to the population in remote areas, malaria was almost wiped out. Unfortunately, today the number of malaria cases in Guyana is increasing. He can be seen on a stamp from Guyana (312).

Fig. 312

Eugène Jamot (1879 – 1937) devoted almost his whole life after finishing his medical studies in Paris to the fight against sleeping sickness. For many African countries, he was the 'conqueror of sleeping sickness' ('vainqueur de la maladie du sommeil'). France and several African countries honoured him with a stamp (186, 187, 188, 191, 313).

Oliveira Gaspar de Vianna (1885 – 1914) (195) worked on Chagas' disease and leishmaniosis in Brazil. He discovered in 1912 that leishmaniosis could be controlled by an intravenous injection with potassium antimonyl tartrate. Unfortunately he died prematurely at the age of 29.

Fig. 313

Saul Aaron Adler (1895 – 1966), born in Russia, studied medicine in the UK. In 1924, he emigrated to Israel and became head of the Department of Parasitology until his retirement in 1965. He is particularly famous for being able to culture *Leishmania* in the lab in wild Syrian hamsters (194).

Pedro Kourí Esmeja (1900 – 1964) was born on Haiti. After having studied physics, chemistry and natural history, he concentrated on medicine at the University of Havana, Cuba, where he became full professor in 1934. He developed methods for diagnosis as well as original treatments for numerous parasitic diseases of man. He wrote a four volume tome on 'Lessons on Parasitology and Tropical Medicine'. He described a new parasite species, the pentastome *Inermicapsifer cubensis*. On his Cuban stamp, attention is focussed on his work on human fasciolosis (209).

Other scientists who made contributions to parasitology

Fig. 314

As mentioned in the last chapter, one may argue that some of the scientists listed below should have been named in Chapter 9. So again, my apologies to those who may disagree with my selection.

Scientists who contributed to our knowledge of malaria

François-Clément Maillot (1804 – 1894) worked in Algeria and contributed to the control of malaria in that country (314).

Joseph Lister (1827 – 1912) is of course well known world wide for his work on sepsis and disinfection. The genus *Listeria* was named after him. However, during his life he also did some work on malaria. Stamps are from the UK (315, 316) and Traanskei (317).

Fig. 315　　　　　　　*Fig. 316*

Fig. 317

Fig. 318

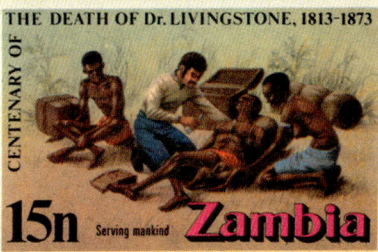

Fig. 319

Scientists who contributed to our knowledge of sleeping sickness

David Livingstone (1813 – 1873) is widely known as an explorer of Africa and for his particular encounter with Stanley (318). During his travels he described sleeping sickness and helped tend the sick (319).

Scientists who contributed to our knowledge of nematode infections

William Osler (1849 – 1919) was a famous Canadian physician (320). He has to be mentioned here because he wrote about verminous pneumonia in a dog. He was honoured with a species, and perhaps a genus name: *Filaroides osleri* designated in some texts as *Oslerus osleri*.

Fig. 320

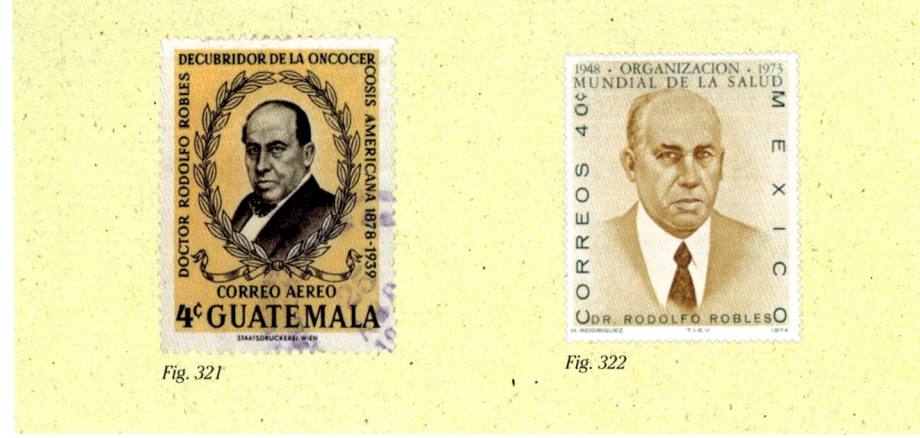

Fig. 321

Fig. 322

Rodolfo Robles (1878 – 1939), born in Guatemala, discovered and described the American form of onchocercosis. Both his home country and Mexico dedicated stamps to him (321, 322).

Victor Manuel Calderon (1889 – 1969) was another scientist from Guatemala who worked on onchocercosis caused by *Onchocerca volvulus* (323).

Scientists who made observations on human cysticercosis

This group consists mainly of pathologists, neurologists or psychiatrists who encountered patients with disorders attributed to tapeworm cysts, the so-called neurocysticercoses (probably metacestode forms of *Taenia solium*). Among them are **Jean-Nicolas Corvisart des Marets** (1755 -1821), known as the personal physician of Napoleon Bonaparte (324). **Jean-Martin Charcot** (1825 – 1893), the French neurologist whose name is associated with studies on hysteria at the Salpêtrière hospital in Paris (325). **Korányi Frigyes** (1828 – 1913) well known for his work on tuberculosis in Hungary (326, 327). **Antoine Depage** (1862 – 1925), the founder of the Red Cross in Belgium who also worked for some time with the famous nurse Edith Cavell (328). **George Ferdinand Isidore Widal** (1862 – 1929), another French physician, and professor in internal anatomy who worked for a year at the Salpêtrière (329). **Korányi Sandor** (1866 – 1944), another Hungarian physician (330). **Karl Schönherr** (1867 – 1943), an Austrian physician, particularly well known as a writer (331). **Egas Moniz** (1874 – 1955), a Portuguese neurologist, but also known for other activities, in his case politics. He was awarded the Nobel prize in Physiology or Medicine in 1949 (332).

Fig. 323

Fig. 324

Fig. 325

Fig. 326

Fig. 327

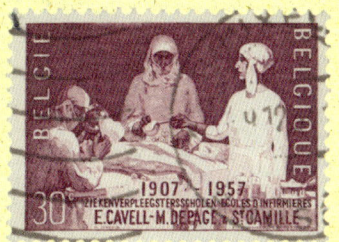

Fig. 328

Fig. 329

Fig. 330

Fig. 331

Fig. 332

CHAPTER 11

Stamps issued to commemorate congresses featuring parasitology

There have been many parasitology congresses. Most have not been important enough to warrant the local postal authorities issuing a stamp. Congress organizers have sometimes been able to arrange a special cancellation, or, today, in some countries it is possible to buy one for use at the congress. In this chapter, we concentrate on the stamps, in chronological order, with the exception of one conference for which the franking mark is the object of particular interest to the true parasitologist.

Fig. 309

1950: Fifth International Congress on Microbiology, held in Brazil. Stamp issued with an image of Oswaldo Cruz (309).

1952: First Congress on Tropical Medicine, held at Lisbon, Portugal. Stamps were issued by Angola (333), Cape Verde Islands (334), Macau (335), Mozambique (336), Portuguese Guinea (337), Portuguese India (not in the author's possession (yet!)), San Tomé and Principe (338) and Timor (339).

1958: Sixth International Congress on Tropical Medicine and Malaria, held at Lisbon, Portugal. Stamps were issued by Portugal (340) and the eight Portuguese colonies. Here we give one example Angola (341), because the stamps are identical with regard to text and differ only with regard to the illustrated plant species.

1964: FAO Congress on Antibrucellosis, held on Malta. Stamp with David Bruce (302)

Fig. 302

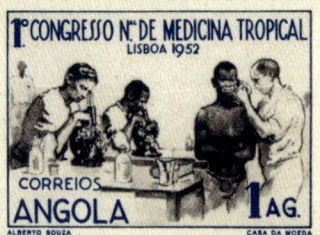

Fig. 333

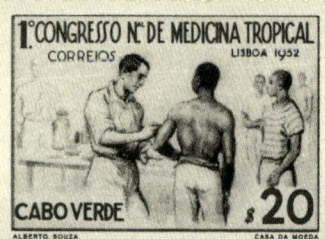

Fig. 334

Fig. 335

Fig. 336

Fig. 337

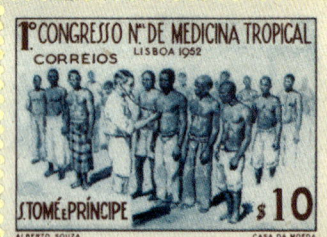

Fig. 338

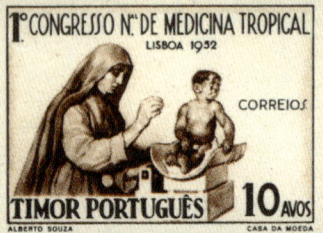

Fig. 339

Fig. 340   Fig. 341

Fig. 342

1968: Eighth International Congress on Tropical Medicine and Malaria, held in Teheran, Iran. Stamp issued by Iran (342)

1969: Third International Congress on Protozoology, held in Leningrad, USSR (now St. Petersburg, Russia). No stamp, but a beautiful cancellation (202)

1975: Third International Conference on Schistosomiasis, held in Cairo, Egypt. Stamp from Egypt (208).

1978: Fourth International Congress on Parasitology (ICOPA IV), held at Warsaw. Stamps issued by Poland (63, 184)

1981: Twelfth International Congress on Hydatidology. Stamp issued by Algeria (204).

1985: Seventh International Congress on Protozoology, held in Nairobi, Kenya. Four stamps were issued (53, 182, 193, 197)

Finally, the International Association for Medical Assistance to Travellers (IAMAT) issued two non-postal stamp labels on malaria and schistosomiasis (343).

Fig. 208

Fig. 202

The last stamp of this book is dedicated not to a parasite, but to an Institution that has contributed so much to our knowledge of parasitology, the CSIRO in Australia.

Fig. 63

Fig. 184

So I would like to dedicate this book not only to the late Peter Nansen, but also to his friend and collaborator, the late Des Hennessy, who worked for many years for the CSIRO (344).

Fig. 204

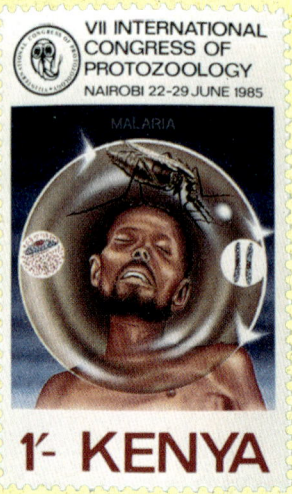

Fig. 53

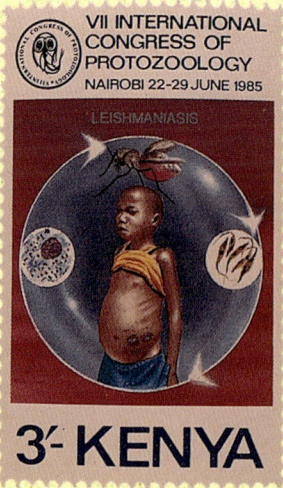

Fig. 193

Fig. 182

Fig. 197

Fig. 343

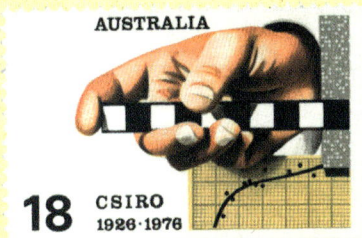

Fig. 344

Bibliography

Borgsteede, F.H.M. 1982. Parasites on stamps. *Parasitology*, 84, LVI.

Borgsteede, F.H.M. 2001. Parasites on stamps: where parasitology meets philately. *Acta Parasitologica* 46, 58-62.

Borgsteede, F.H.M. 2002. Parasieten op postzegels. *Filatelie* 2002/02, 114-116.

Borgsteede, F.H.M. 2004. Parasitology world philatelic exhibit. Abstract book EMOP VIII, Valencia, Spain, 18-23 July 2004, p. 20.

Doby, J.-M. 1981. Paludisme et timbres-poste. In: *Parasitological topics. A presentation volume to P.C.C. Garnham FRS on the occasion of his 80th birthday* (Ed. E.U. Canning). Published by the Society of Protozoologists, London, 86-93.

Doby, J.-M. 1986. Quand la philatélie se pique de palu..... *Histoire de la Médecine*, 194, 22-26.

Enigk, K. 1986. Geschichte der Helminthologie im deutschsprachigen Raum. Gustav Fisher Verlag, Stuttgart, New York.

Grieshop, J.I. 1990. Licking pests. *American Entomologist*, 36, 283-286.

Guarda, F. 2005. Piccola storia della medicina veterinaria raccontata dai francobolli. Edito cura della *Fondazione initiative zooprofilattiche e zootecniche*-Brescia 58, 183 pp.

Guarda, F., Rossotti, R. 1998. Sull'Arca con Noé. Gli affascinanti misteri degli animali e il mondo veterinario nei francobolli. Ed. Christiano Giraldi, Ozzano dell'Emilia, Italy.

Gysin, P., Locke D.R. 2004. Veterinary Science on Postal Cancellations/ Veterinärmedizin auf Poststempeln. 2nd Ed. ISBN 0-9547692-0-1.

Hamel, D.R. 1990. Insects on stamps. *American Entomologist*, 36, 273-281.

Kassai, T., Burt, M.D.B. 1994. A plea for consistency. *Parasitology Today*, 10, 127-128.

Schwarz, A.W. 1996. Veterinary philately. Issued by the author and dedicated to Donfilex. Utrecht, the Netherlands, 144 pp.